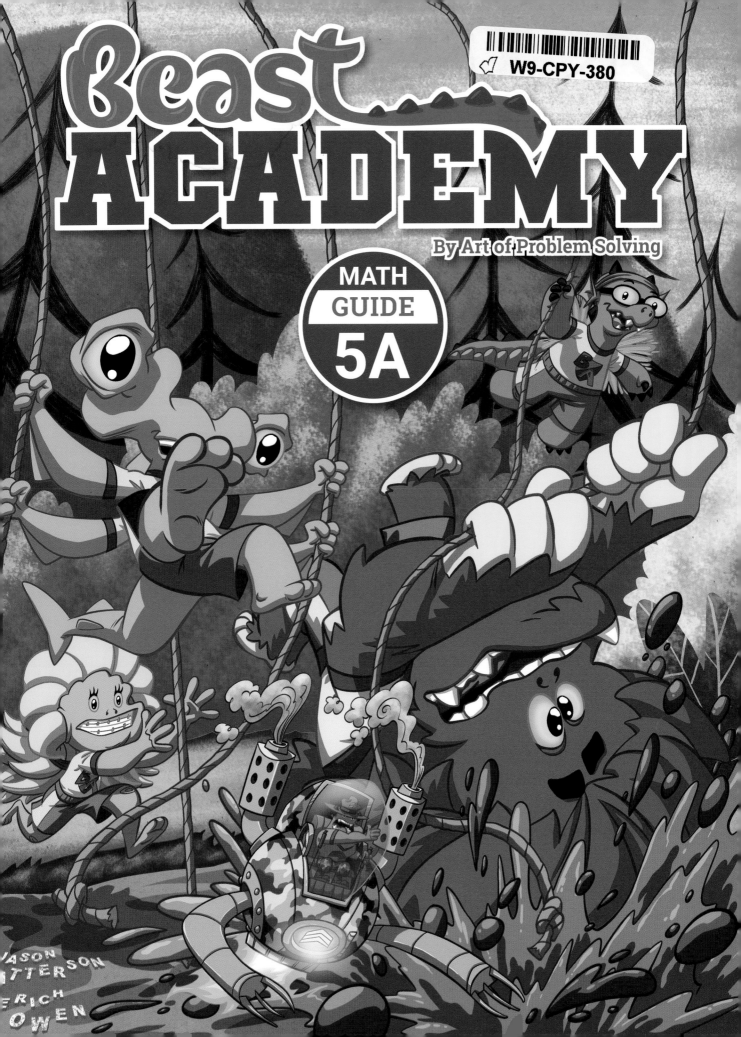

Published by: AoPS Incorporated
                     10865 Rancho Bernardo Rd Ste 100
                     San Diego, CA 92127-2102
                     info@BeastAcademy.com

ISBN: 978-1-934124-60-4

Beast Academy is a registered
trademark of AoPS Incorporated.

Written by Jason Batterson and Kyle Guillet
Illustrated by Erich Owen
Additional Illustrations by Paul Cox
Colored by Greta Selman

Visit the Beast Academy website at www.BeastAcademy.com.
Visit the Art of Problem Solving website at www.artofproblemsolving.com.
Printed in the United States of America.
2017 Printing.

Become a Math Beast!
For additional books,
printables, and more, visit
BeastAcademy.com

This is Guide 5A in a four-book series:

Contents:

Chapter 3: Expressions & Equations

Lizzie
"The BOOkwOrm"
can name every
dragOn species
On
Beast Island
(alphabetically)

Alex
"The Executive"
Plans tO run fOr
city cOmptrOller
when he's Old enOugh
fOr public Office

Winnie
"The Firecracker"
Feisty!
GrOws 50 times
her Original size when angry!
(nOt really, but it's fun
to draw her that way)

GrOgg (me)
"The ^least common DenOminatOr"

ALter EgO:
FractiOn JacksOn!

Mr. Wriggles

kraken
Shop Teacher

Favorite pattern?
Arrrrrgyle

Favorite holiday?
ArrrrbOr Day

Favorite element?
~~Arrrrrgon~~
Gold

Fiona
Math Team Coach

Donated her hair to
"Braids for Mermaids"
this summer

"Calamitous Clod"

Professor Grok
Math Lab
(full of booby traps)

Constantly
captured by

Ms. Q.
Math Teacher

Spends a lot
of time
with Mr. A.

R&G
campus Maintenance
Engineer(s?)

Let me ride
in their
golf cart
once!

Sgt. Rote →
Gym Teacher

Can bench press
three times
his own bodyweight!
(4 lbs.)

The Headmaster
How to use this book

Welcome to Beast Academy!

This book is called the Guide.

There is also a separate Practice book with lots of problems you can use to sharpen your skills.

The Guide is written like a comic book.

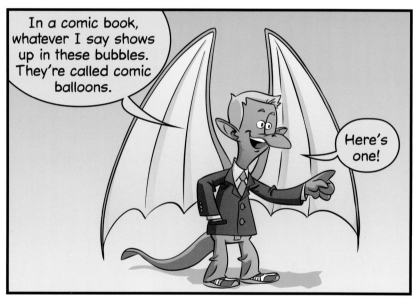

In a comic book, whatever I say shows up in these bubbles. They're called comic balloons.

Here's one!

Each character has a different balloon color. This makes it easy to tell who is talking.

My balloons are purple!

The story is told in panels.

Panels usually have a rectangular frame around them...

...like this one.

Contents: Chapter 1

See page 6 in the Practice book for a recommended reading/practice sequence for Chapter 1.

Chapter 1:
3D Solids

Salvador
Yetí
Persistence
of Elefinches

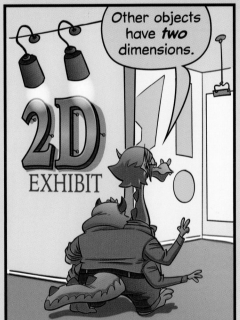

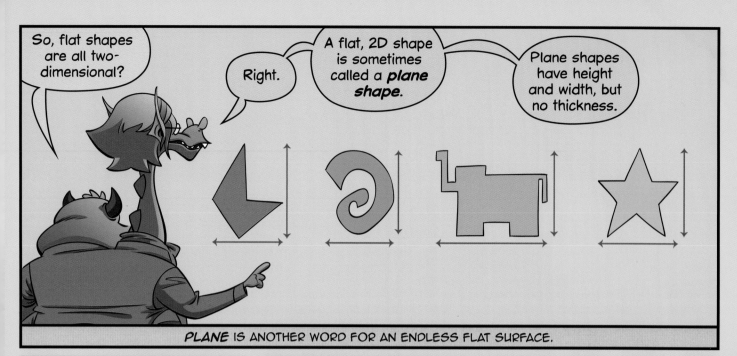

So, flat shapes are all two-dimensional?

Right.

A flat, 2D shape is sometimes called a *plane shape*.

Plane shapes have height and width, but no thickness.

PLANE IS ANOTHER WORD FOR AN ENDLESS FLAT SURFACE.

I see. So, a *3D* object has *three* dimensions.

3D EXHIBIT

Yep. Let's check out the 3D exhibit!

Cooool.

All of these pieces are three-dimensional *solids*.

They take up space in three dimensions!

G*Y*M
SOLIDS

Line up, polliwogs!

Today's physical training will help you improve your balance, stamina, and coordination.

In front of you is an obstacle course made of prisms, pyramids, and various other *geometric solids*.

???

Some of you have not learned to identify the major groups of three-dimensional solids.

Who can tell me which one of these obstacles is a rectangular prism?

Sir, the big blue one, sir.

Very good, Tinkerbell.

What makes that solid a rectangular prism?

Sir, a **prism** is a solid with two congruent **faces** that are parallel.

These congruent faces are called the **bases** of the prism.

We name a prism by the shape of its bases.

The bases of this prism are rectangles, so it is a rectangular prism, sir!

THE FLAT SIDES OF A GEOMETRIC SOLID ARE ITS *FACES*. TWO FACES ARE *PARALLEL* IF THEY ARE IN PLANES THAT NEVER CROSS.

Excellent. You will begin my obstacle course by climbing up one face of the rectangular prism.

After climbing down the opposite face, you will continue to this set of six smaller prisms.

Hop from the triangular prism...

...and finally, onto the octagonal prism.

...to the rectangular prism...

...to the pentagonal prism...

...to the hexagonal prism...

...to the heptagonal prism...

I see. This octagonal prism has an octagon on top...

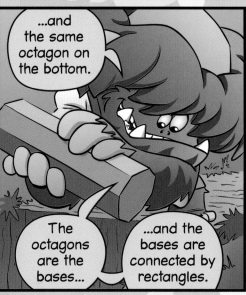

...and the same octagon on the bottom.

The octagons are the bases...

...and the bases are connected by rectangles.

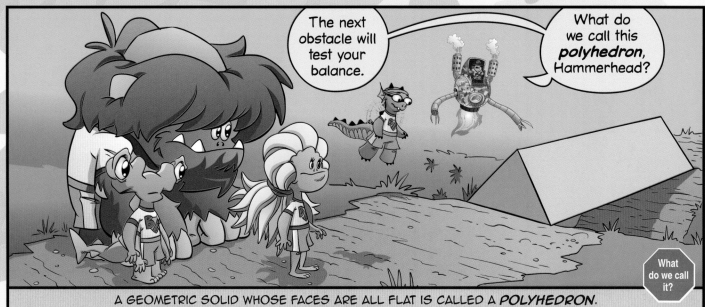

The next obstacle will test your balance.

What do we call this *polyhedron*, Hammerhead?

What do we call it?

A GEOMETRIC SOLID WHOSE FACES ARE ALL FLAT IS CALLED A *POLYHEDRON*.

20

Sir, it's a triangular prism, sir!

Excellent. Regardless of which face is on the bottom...

...a triangular prism is still a triangular prism, and its bases are still triangles.

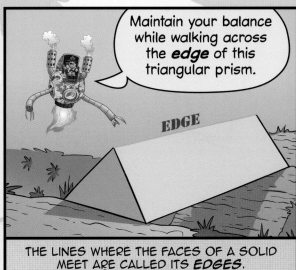

Maintain your balance while walking across the *edge* of this triangular prism.

EDGE

THE LINES WHERE THE FACES OF A SOLID MEET ARE CALLED ITS *EDGES*.

Next, you must cross this moat using these five pyramids.

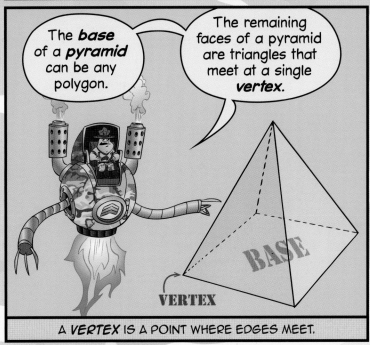

The *base* of a *pyramid* can be any polygon.

The remaining faces of a pyramid are triangles that meet at a single *vertex*.

BASE

VERTEX

A *VERTEX* IS A POINT WHERE EDGES MEET.

The vertex where all of the triangular faces meet is called the *apex* of the pyramid.

APEX

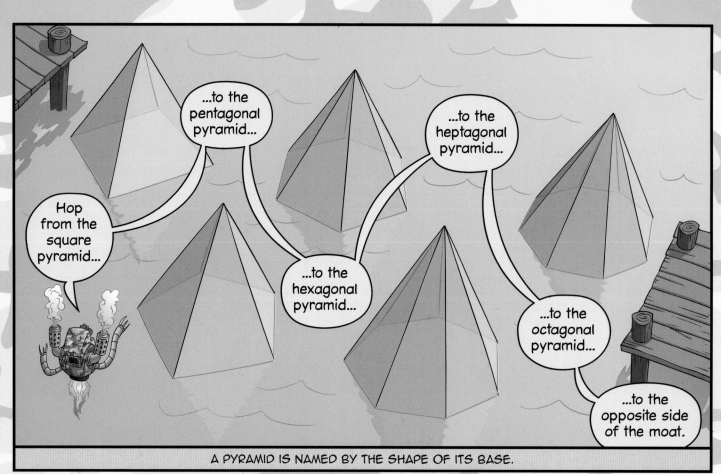

A PYRAMID IS NAMED BY THE SHAPE OF ITS BASE.

3D Solids

Prism:

Has two congruent parallel faces.
Other faces (lateral faces) are all rectangles.
Prisms are named by the shape of their bases.

*All of the prisms in this book are called right prisms and have lateral faces that are rectangles. But, prisms can have lateral faces that are parallelograms.

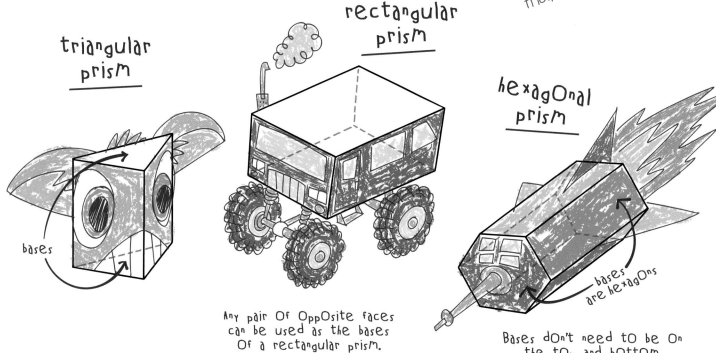

triangular prism

bases

rectangular prism

Any pair of opposite faces can be used as the bases of a rectangular prism.

hexagonal prism

bases are hexagons

Bases don't need to be on the top and bottom.

Pyramid:

Any polygon can be the base of a pyramid.
All the other faces are triangles that meet at a vertex.

square pyramid

triangular pyramid

(Any face of a triangular pyramid can be used as the base.)

octagonal pyramid

cylinder:
Like a prism, but with a circular base.

sphere:
A ball. Every point on its surface is the same distance from its center.

cone:
Like a pyramid, but with a circular base.

Five Platonic Solids:
All regular faces. All faces congruent. same # of faces meet at each vertex.

icosahedron
20 equilateral triangle faces

bump!

what's up, cube!

cube
6 square faces

tetrahedron
4 equilateral triangle faces

octahedron
8 equilateral triangle faces

dodecahedron
12 regular pentagon faces

A CUBE IS A SPECIAL TYPE OF RECTANGULAR PRISM IN WHICH ALL SIX FACES ARE SQUARES.

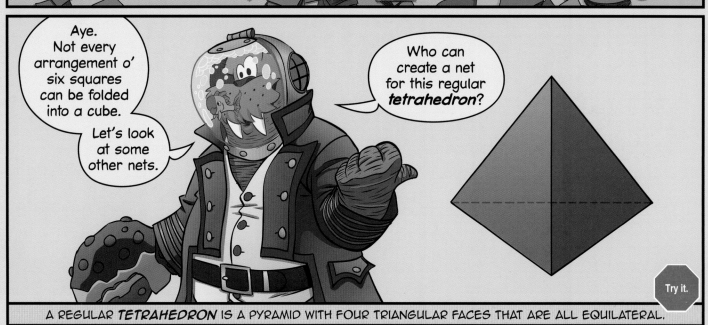

A REGULAR **TETRAHEDRON** IS A PYRAMID WITH FOUR TRIANGULAR FACES THAT ARE ALL EQUILATERAL.

The four faces of a regular tetrahedron are equilateral triangles.

If we arrange them like this, we can fold the three outer triangles up...

...to meet at the apex.

Be there another way to create a net for a regular tetrahedron?

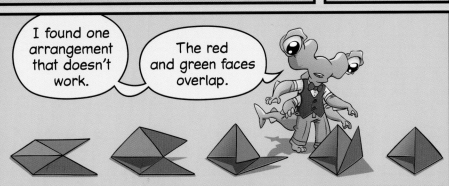

I found one arrangement that doesn't work.

The red and green faces overlap.

But, if we arrange four triangles like this...

...they fold into a tetrahedron.

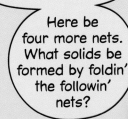

Excellent foldin'.

Here be four more nets. What solids be formed by foldin' the followin' nets?

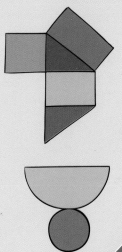

What solids are formed by folding these nets?

29

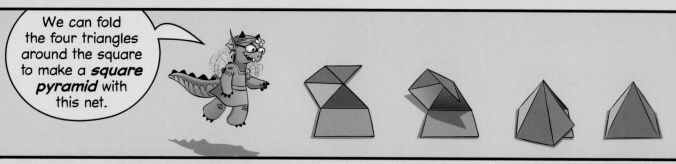

We can fold the four triangles around the square to make a **square pyramid** with this net.

This net folds into a wedge, which is a **triangular prism**.

We can wrap the rectangle into a tube and put circles on top and bottom to make a **cylinder**.

And for this net, we can wrap up the half-circle and place it on top of the full circle to make a **cone**!

Ye be a clever bunch.

Now let's build some bigger solids.

The *surface area* of a solid is the total area of all its faces.

Since all six faces of a cube are the same, a cube's surface area is six times the area of one of its faces.

Let's try a slightly more difficult solid.

3 in
5 in
8 in

What's the surface area of this prism?

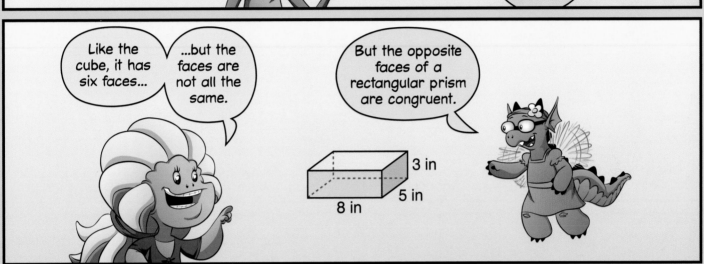

Like the cube, it has six faces...

...but the faces are not all the same.

But the opposite faces of a rectangular prism are congruent.

3 in
5 in
8 in

The top and bottom are both 5-by-8.

3
5
8

The left and right are both 3-by-5.

3
5
8

And the front and back are both 3-by-8.

3
5
8

Surfaces	Area	
Top & Bottom	5×8	×2
Left & Right	3×5	×2
Front & Back	3×8	×2

That's all six faces.

Compute the total surface area of the prism.

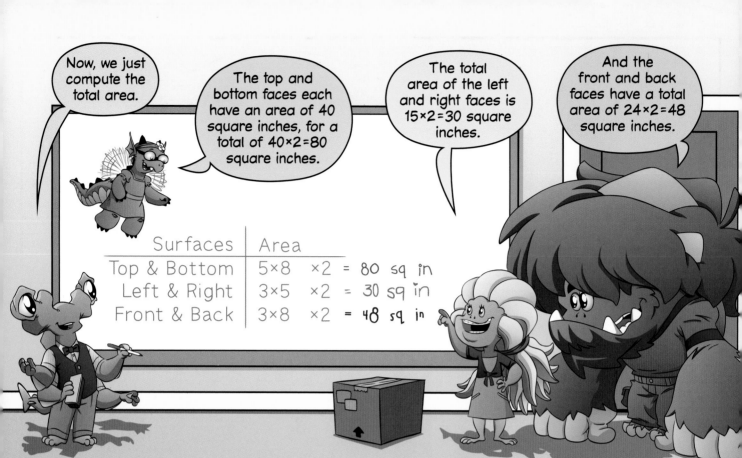

Now, we just compute the total area.

The top and bottom faces each have an area of 40 square inches, for a total of 40×2=80 square inches.

The total area of the left and right faces is 15×2=30 square inches.

And the front and back faces have a total area of 24×2=48 square inches.

Surfaces	Area			
Top & Bottom	5×8	×2	=	80 sq in
Left & Right	3×5	×2	=	30 sq in
Front & Back	3×8	×2	=	48 sq in

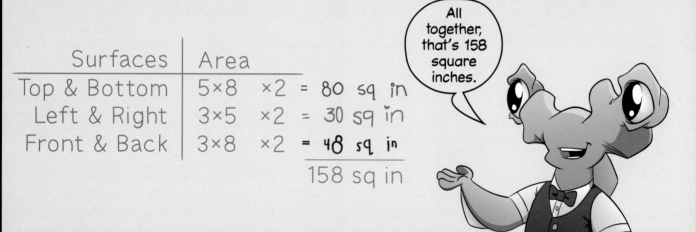

Surfaces	Area			
Top & Bottom	5×8	×2	=	80 sq in
Left & Right	3×5	×2	=	30 sq in
Front & Back	3×8	×2	=	48 sq in
				158 sq in

All together, that's 158 square inches.

Great work! Finding the surface area of solids is really all about staying organized.

This becomes even more important when finding the surface area of a triangular prism.

How would you compute the surface area of this prism?

10

8

20

6

Try it.

34

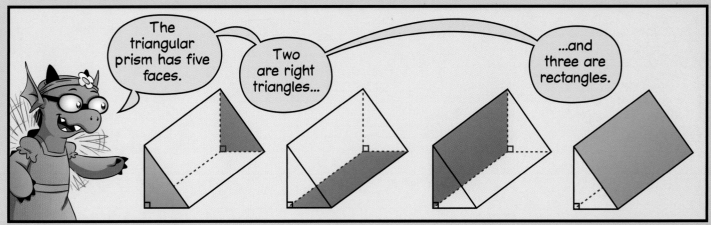

The triangular prism has five faces.

Two are right triangles...

...and three are rectangles.

We need to find the areas of all five faces and add them together.

Let's start by finding the areas of the triangles.

To find the area of a triangle, we multiply the length of its base by its height and divide by two.

Each triangle has an area of 24 square units.

Area
$= 6 \times 8 \div 2$
$= 24$

REVIEW TRIANGLE AREA IN GUIDE 3D.

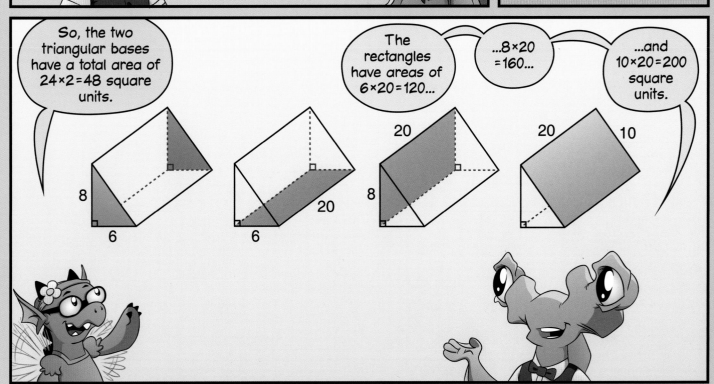

So, the two triangular bases have a total area of 24×2=48 square units.

The rectangles have areas of 6×20=120...

...8×20 =160...

...and 10×20=200 square units.

So, the combined area of all five faces is 48+120+160+200=528 square units.

Well done!

Let's try one last surface area problem.

Each of these four cubes has a surface area of exactly 5 square inches.

By attaching each of the blue cubes to a face of the pink cube...

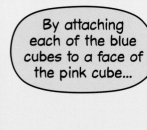

...I can make this solid.

What is its surface area?

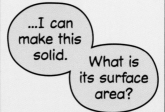

If we find the area of one cube face...

...we can multiply that by the number of faces that are showing.

The total surface area of 6 cube faces is 5 square inches.

So, each face has an area of $5 \div 6 = \frac{5}{6}$ square inches.

Great! How many of the cube faces are showing?

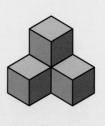

How many do you count?

36

If each face has an area of $\frac{5}{6}$...

...and there are 18 faces...

...then the total area of the solid is $18 \times \frac{5}{6} = 15$ square inches.

$$18 \times \frac{5}{6} = \frac{18 \times 5}{6} = \frac{18}{6} \times 5 = 3 \times 5 = 15$$

We can also solve the problem without ever multiplying fractions.

If a cube with **6** square faces has a surface area of 5 square inches...

...then a solid with **18** of the same square faces will have three times the surface area.

Three times 5 square inches is 15 square inches.

Nice work! Now, let's see what's in this box.

Practice: Pages 14-31

Our new Math Team shirts!

Uh oh.

I think that someone misjudged my surface area.

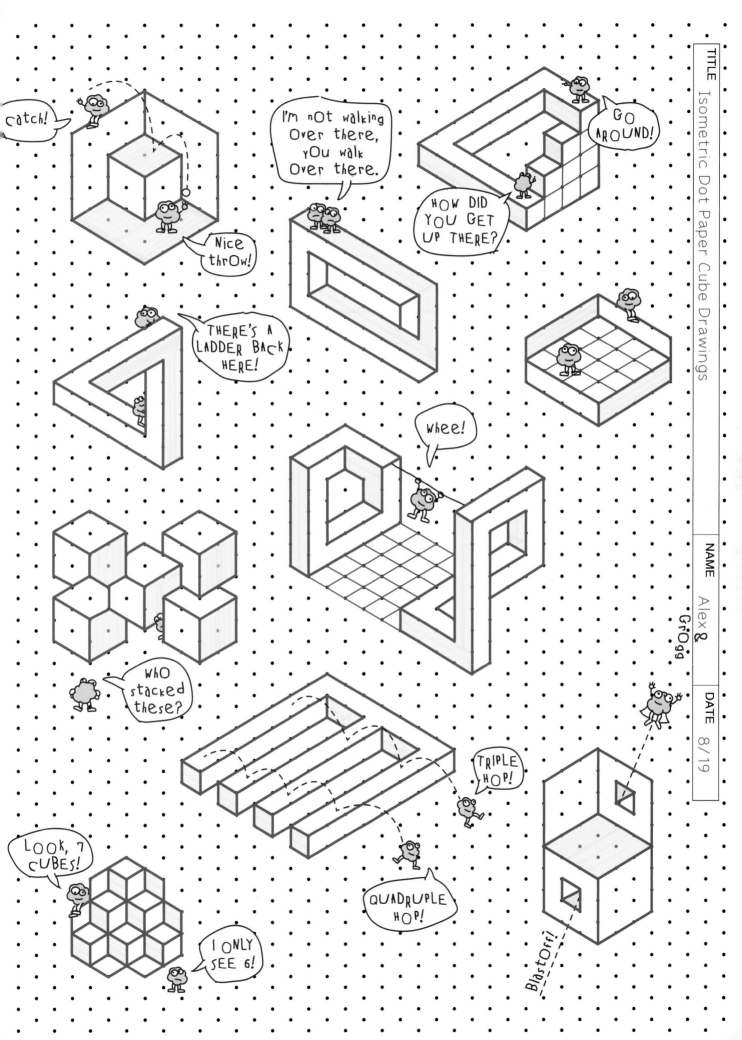

Ms. Q. Volume

What is volume?

Volume is the amount of space a solid fills.

How do we measure volume?

With cubes.

What do you mean, Alex?

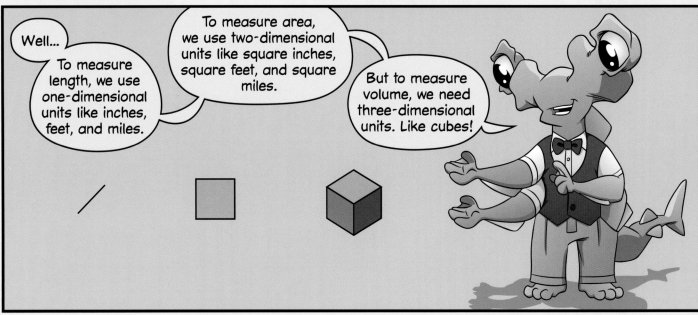

Well...

To measure length, we use one-dimensional units like inches, feet, and miles.

To measure area, we use two-dimensional units like square inches, square feet, and square miles.

But to measure volume, we need three-dimensional units. Like cubes!

Very good, Alex.

Volume is measured in cubic units...

...like cubic inches, cubic feet, and cubic miles.

How could we find the volume of this rectangular prism in cubic centimeters?

5 cm

4 cm

3 cm

WE ALSO LEARNED TO MEASURE VOLUME USING OUNCES, GALLONS, AND LITERS IN GUIDE 3C.

What is the volume of this prism?

A cube with an edge length of 1 centimeter has a volume of 1 cubic centimeter.

We need to figure out how many of these cubes it will take to fill the prism.

1 cm

1 cubic cm

The base of the prism is 4 cm by 3 cm. We can cover the base using 4×3=12 cubes.

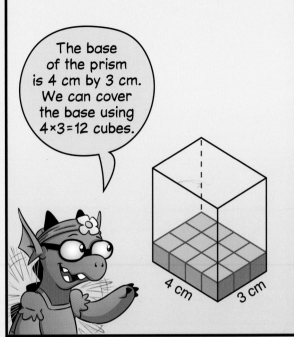

4 cm   3 cm

4 cm   3 cm

4 cm   3 cm

Area: 4×3 = 12 Sq Cm

The number of cubes it takes to cover the base is the same as the area of the base.

The prism is 5 centimeters high. So, we can stack 5 layers of 12 cubes.

5 layers of 12 cubes makes a total of 5×12=60 cubic centimeters.

5 cm

4 cm   3 cm

5×12=60 cubic cm

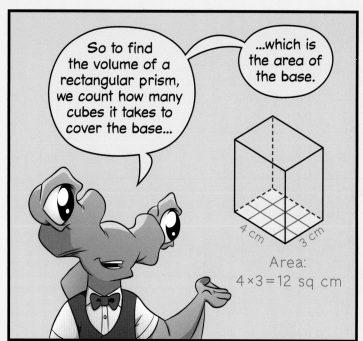

So to find the volume of a rectangular prism, we count how many cubes it takes to cover the base...

...which is the area of the base.

4 cm   3 cm

Area:
4 × 3 = 12 sq cm

And we multiply by the number of layers we stack to make the prism...

...which is the prism's height.

5 cm

Volume:
12 × 5 = 60 cu cm

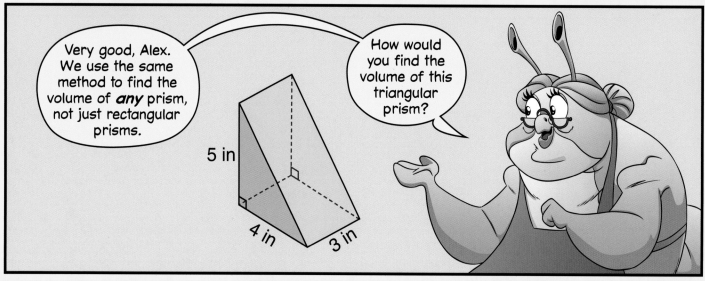

Very good, Alex. We use the same method to find the volume of *any* prism, not just rectangular prisms.

How would you find the volume of this triangular prism?

5 in

4 in   3 in

First we need to know how many cubes it takes to cover the base of the prism.

Wait, which face is the base?

It's a *triangular* prism. Its bases are *triangles*.

5 in

4 in   3 in

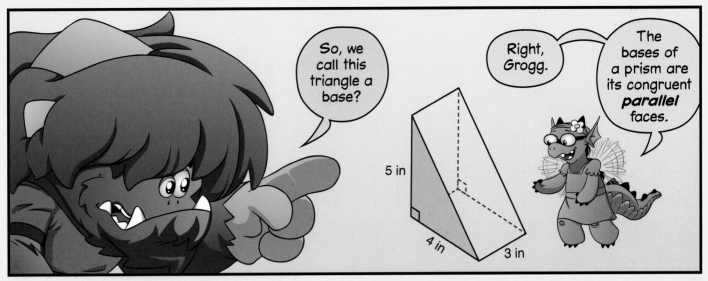

So, we call this triangle a base?

Right, Grogg.

The bases of a prism are its congruent *parallel* faces.

Let's flip it so that the base is on the bottom.

How many cubes does it take to cover the base?

When we try to cover a triangular base with cubes, some stick out.

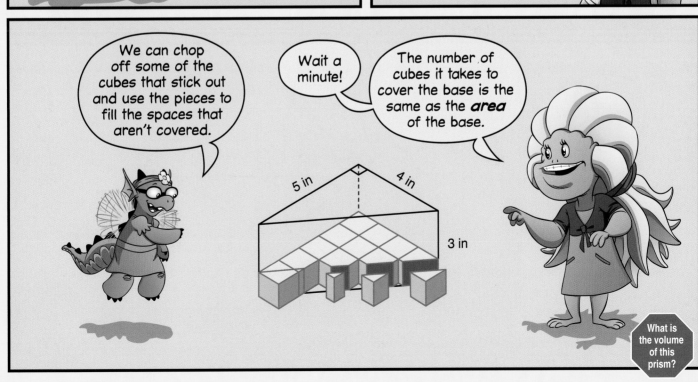

We can chop off some of the cubes that stick out and use the pieces to fill the spaces that aren't covered.

Wait a minute!

The number of cubes it takes to cover the base is the same as the *area* of the base.

What is the volume of this prism?

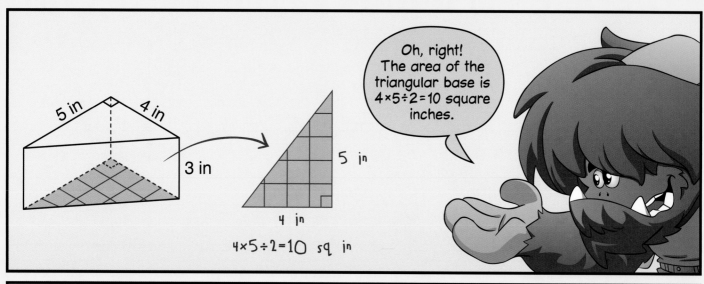

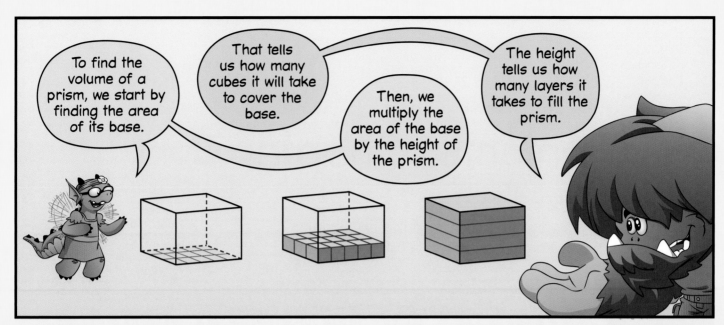

To find the volume of a prism, we start by finding the area of its base.

That tells us how many cubes it will take to cover the base.

Then, we multiply the area of the base by the height of the prism.

The height tells us how many layers it takes to fill the prism.

We can use a big B for the area of the prism's base.

And an h for the prism's height.

The volume of a prism is the area of its base times its height. $V = B \times h$.

Volume = $B \times h$

Nicely done, little monsters.

What in the world are you doing, Grogg?

I'm trying to measure my volume...

...in cheese cubes!

45

Practice: Pages 32-39

Contents: Chapter 2

See page 40 in the Practice book for a recommended reading/practice sequence for Chapter 2.

Chapter 2:
Integers

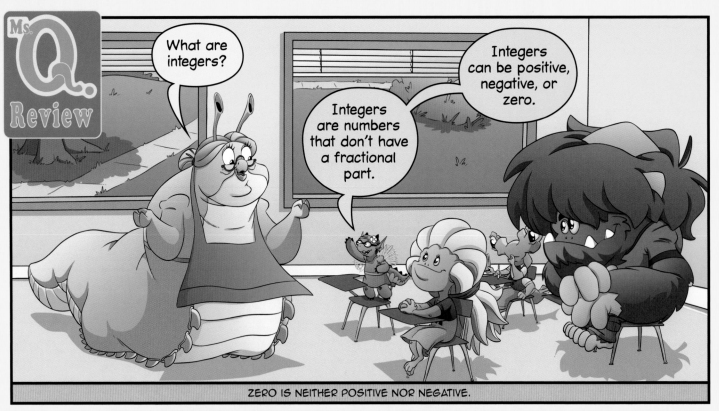

ZERO IS NEITHER POSITIVE NOR NEGATIVE.

2+3

−4+7

−19+(−32)

12+(−27)

Try all four.

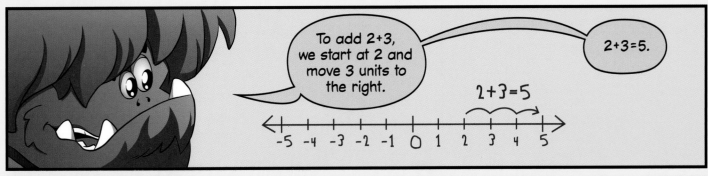

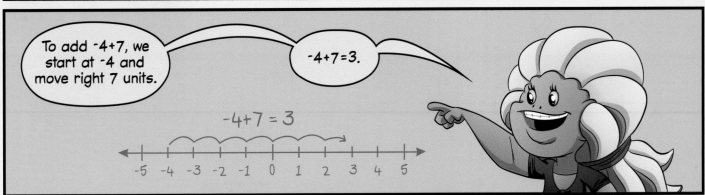

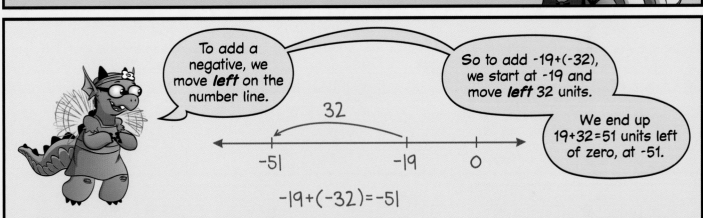

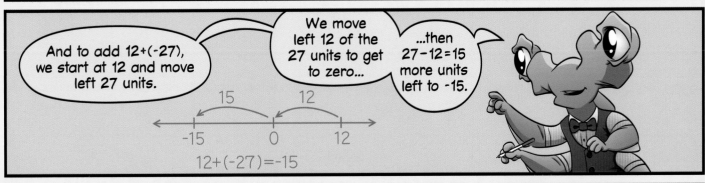

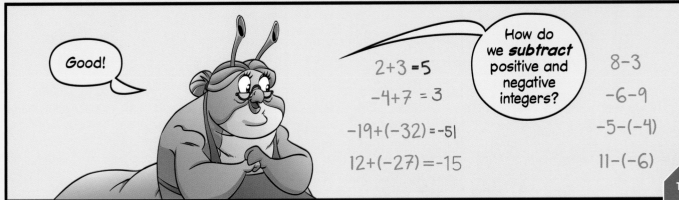

Try all four.

To subtract an integer, we add its opposite.

What does Lizzie mean by **opposite?**

Two integers are opposites if their sum is zero.

For example, the opposite of *positive* 3 is *negative* 3, because $3+(-3)=0$.

We can rewrite all subtraction as addition.

$$8-3 = 8+(-3)$$
$$-6-9 = -6+(-9)$$
$$-5-(-4) = -5+4$$
$$11-(-6) = 11+6$$

Subtracting a *negative* 6 is the same as adding a *positive* 6.

$$8-3 = 8+(-3)$$
$$= 5$$

$$-6-9 = -6+(-9)$$
$$= -15$$

Sometimes, it's best to leave subtraction as subtraction. $8-3=5$.

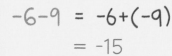

-6-9 =-15.

Who's the new student?

$$-5-(-4) = -5+4$$
$$= -1$$

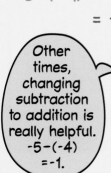

Other times, changing subtraction to addition is really helpful. $-5-(-4)=-1$.

$$11-(-6) = 11+6$$
$$= 17$$

And $11-(-6)=17$.

Practice: Pages 41-43

MULTIPLYING INTEGERS

Captain Kraken, who are all these pirates?

Arrr... These be some o' the greatest beast captains in pirate history.

Here be Captain Kelpler...

...Kelpler charted the entire coastline o' Beast Island.

He was also one o' the island's earliest experts in integer operations.

CAPT. KELPLER

'Twas Kelpler who first pondered the result o' multiplyin' a positive by a negative.

For example, Kelpler correctly deduced the product $4 \times (-3)$.

Whoa. How did he do that?

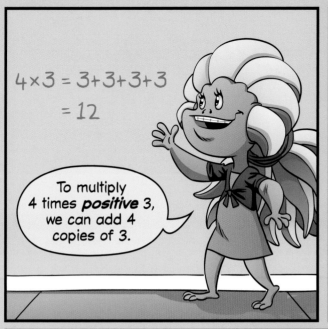

$4 \times 3 = 3+3+3+3$

$= 12$

To multiply 4 times **positive** 3, we can add 4 copies of 3.

$4 \times (-3) = (-3)+(-3)+(-3)+(-3)$

$= -12$

So to multiply 4 times **negative** 3, we should add 4 copies of -3!

$4 \times (-3) = -12!$

Aye. Excellent figurin', little monsters.

After convincin' his crew that $4 \times (-3)$ be -12, Kelpler set out to compute even more products.

Try each o' these.

$3 \times (-7)$

$(-5) \times 9$

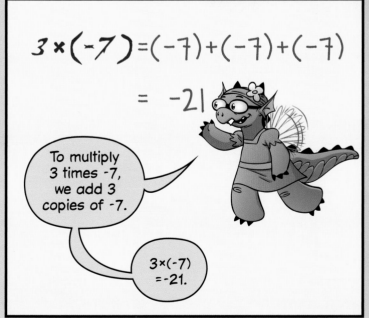

$3 \times (-7) = (-7)+(-7)+(-7)$

$= -21$

To multiply 3 times -7, we add 3 copies of -7.

$3 \times (-7) = -21.$

To multiply -5 times 9, we add -5 copies of 9...

Scratch Scratch

Wait, ummm...

$(-5) \times 9 =$

How could you multiply $(-5) \times 9$?

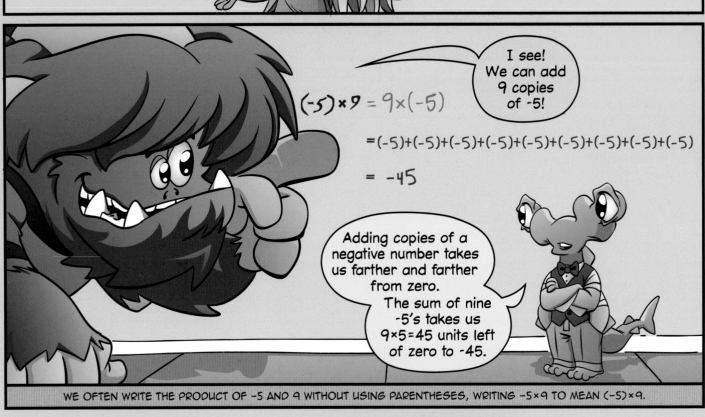

WE OFTEN WRITE THE PRODUCT OF −5 AND 9 WITHOUT USING PARENTHESES, WRITING −5×9 TO MEAN (−5)×9.

We can write -19×36 as 36×(-19).

Then, adding 36 copies of -19 takes us 36×19=684 units left of zero...

...to -684!

$$-19 \times 36 = 36 \times (-19)$$
$$= -684$$

Since adding copies of a negative number takes us farther and farther from zero...

...the product of a positive number and a negative number is always *negative*!

Aye!

Captain Kelpler was able to show that the product of a positive and a negative always be negative.

$$4 \times (-3) = -12$$
$$3 \times (-7) = -21$$
$$(-5) \times 9 = -45$$
$$-19 \times 36 = -684$$

What if we multiply *two* negatives?

'Twas Captain Kelpler's nephew, Captain Krill, who pondered this very question.

Captain Krill began by computin' the followin' list o' products.

$$4 \times (-3) =$$
$$3 \times (-3) =$$
$$2 \times (-3) =$$
$$1 \times (-3) =$$
$$0 \times (-3) =$$
$$-1 \times (-3) =$$
$$-2 \times (-3) =$$

Compute each product in the list.

$4 \times (-3) = -12$
$3 \times (-3) = -9$
$2 \times (-3) = -6$
$1 \times (-3) = -3$
$0 \times (-3) = 0$
$-1 \times (-3) =$
$-2 \times (-3) =$

We've learned how to compute the first five products.

I'm not sure about -1×(-3).

$4 \times (-3) = -12$
$3 \times (-3) = -9$ +3
$2 \times (-3) = -6$ +3
$1 \times (-3) = -3$ +3
$0 \times (-3) = 0$ +3
$-1 \times (-3) = 3$ +3
$-2 \times (-3) = 6$ +3

If we follow the pattern, we get -1×(-3)=3...

...and -2×(-3) =6.

In this case, when we multiply two negatives, we get a positive.

Is that always true?

Aye. Multiplyin' two negatives always gives a positive result.

SEE WINNIE'S HOMEWORK ON PAGE 59 FOR ANOTHER EXPLANATION OF WHY THIS IS TRUE.

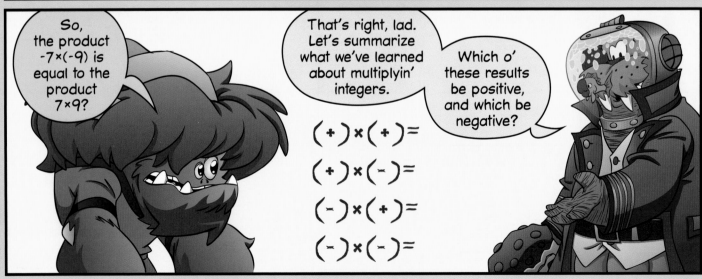

So, the product -7×(-9) is equal to the product 7×9?

That's right, lad. Let's summarize what we've learned about multiplyin' integers.

Which o' these results be positive, and which be negative?

$(+) \times (+) =$

$(+) \times (-) =$

$(-) \times (+) =$

$(-) \times (-) =$

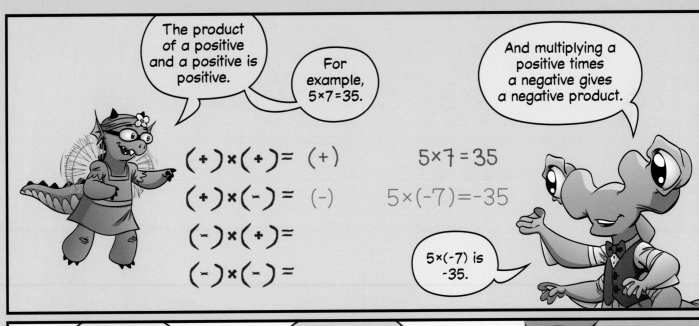

The product of a positive and a positive is positive.

For example, 5×7=35.

And multiplying a positive times a negative gives a negative product.

$(+)\times(+)=(+)$

$(+)\times(-)=(-)$

$(-)\times(+)=$

$(-)\times(-)=$

$5\times7=35$

$5\times(-7)=-35$

5×(-7) is -35.

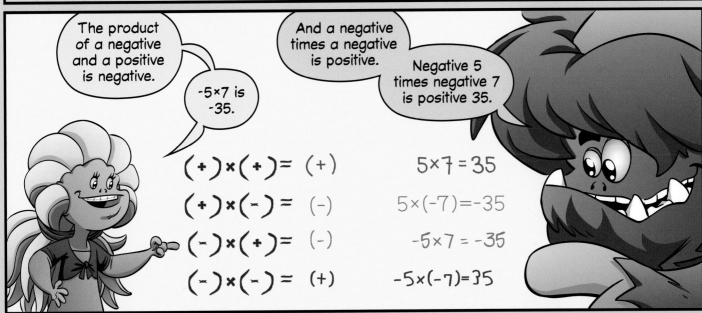

The product of a negative and a positive is negative.

-5×7 is -35.

And a negative times a negative is positive.

Negative 5 times negative 7 is positive 35.

$(+)\times(+)=(+)$

$(+)\times(-)=(-)$

$(-)\times(+)=(-)$

$(-)\times(-)=(+)$

$5\times7=35$

$5\times(-7)=-35$

$-5\times7=-35$

$-5\times(-7)=35$

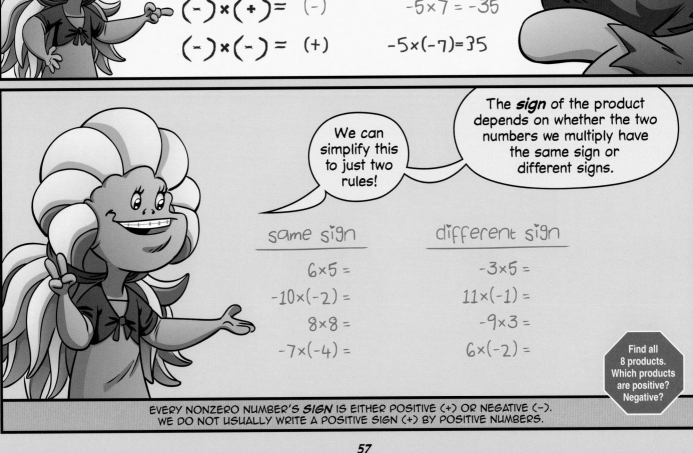

We can simplify this to just two rules!

The **sign** of the product depends on whether the two numbers we multiply have the same sign or different signs.

same sign

$6\times5=$

$-10\times(-2)=$

$8\times8=$

$-7\times(-4)=$

different sign

$-3\times5=$

$11\times(-1)=$

$-9\times3=$

$6\times(-2)=$

Find all 8 products. Which products are positive? Negative?

EVERY NONZERO NUMBER'S **SIGN** IS EITHER POSITIVE (+) OR NEGATIVE (–).
WE DO NOT USUALLY WRITE A POSITIVE SIGN (+) BY POSITIVE NUMBERS.

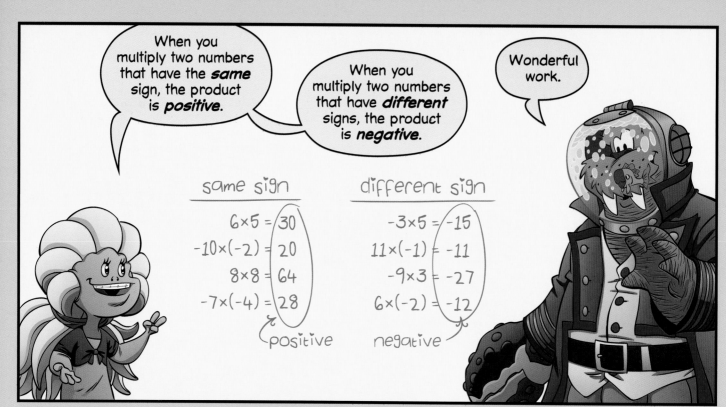

PRACTICE | Solve each.

1. Compute: $5 \times (-4+4)$

$$= 5 \times 0$$
$$= 0$$

1. ____0____

2. Use the distributive property to rewrite the expression $5 \times (-4+4)$ as the sum of two products.

2. $\underline{5 \times (-4) + 5 \times 4}$

3. ★ 🖉 Use your answers to problems 1 and 2 to compute the value of $5 \times (-4)$. Explain your reasoning.

$$5 \times (-4+4) = 0$$
$$5 \times (-4) + 5 \times 4 = 0$$
$$\underline{5 \times (-4) + 20} = 0$$

Two numbers that have a sum of zero are opposites.
So, $5 \times (-4)$ is the opposite of $5 \times 4 = 20$.
That means $5 \times (-4) = \boxed{-20.}$

4. ★ 🖉 Use the reasoning from problem 3 above to show that $-4 \times (-5) = 20$.

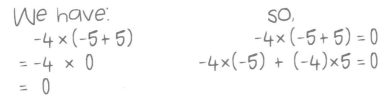

We have:
$$-4 \times (-5+5)$$
$$= -4 \times 0$$
$$= 0$$

So,
$$-4 \times (-5+5) = 0$$
$$-4 \times (-5) + (-4) \times 5 = 0$$

Two numbers that have a sum of zero are opposites.
So, $-4 \times (-5)$ and $(-4) \times 5$ are opposites.
We showed in problem 3 that $5 \times (-4) = -20$, so $(-4) \times 5 = -20$.
$-4 \times (-5)$ is the opposite of $(-4) \times 5$, so $-4 \times (-5) = \boxed{20.}$

Have you done your math homework yet?

Yep. It should be right he...

...uhh...

...uh-oh.

I finished all but the last problem.

10. Compute $(-1)^{99}$.

That one was fun.

How did you solve it?

First I tried to multiply ninety-nine -1's.

That doesn't sound like fun.

It wasn't, but once I got to $(-1)^5$, I noticed a pattern that made it easy.

What is $(-1)^{99}$?

Multiplying by -1 changes a negative to positive...

...or a positive to negative.

I see. It all depends on whether the exponent is odd or even!

$(-1)^1 = \boxed{-1}$

$(-1)^2 = (-1) \times (-1) = \boxed{1}$

$(-1)^3 = (-1)^2 \times (-1) = 1 \times (-1) = \boxed{-1}$

$(-1)^4 = (-1)^3 \times (-1) = -1 \times (-1) = \boxed{1}$

$(-1)^5 = (-1)^4 \times (-1) = 1 \times (-1) = \boxed{-1}$

If the exponent is odd, we get -1.

Exactly. And if the exponent is even, we get positive 1!

$(-1)^1 = -1$

$(-1)^2 = 1$

$(-1)^3 = -1$

$(-1)^4 = 1$

$\vdots$

$(-1)^{99} = -1$

Since 99 is odd, $(-1)^{99}$ is -1.

Cool!

That was actually the *first* problem I did, and it made all the other problems easier.

What do you mean?

Look at problem 7, for example.

I got -2,880.

Nope. The answer has to be positive.

7. Evaluate $2 \times 3 \times (-4) \times (-5) \times (-4) \times 3 \times (-2)$.

Is Grogg right?

61

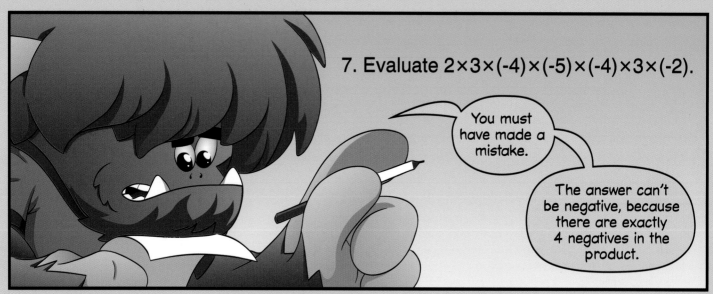

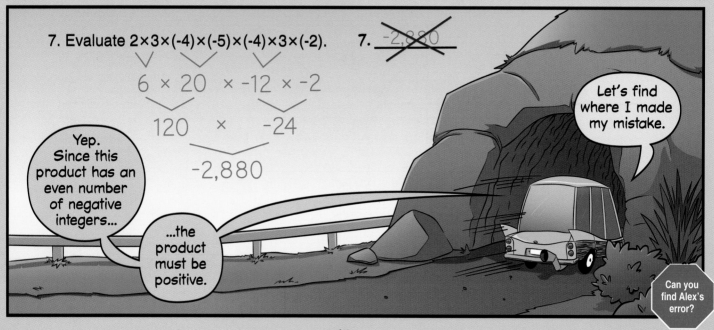

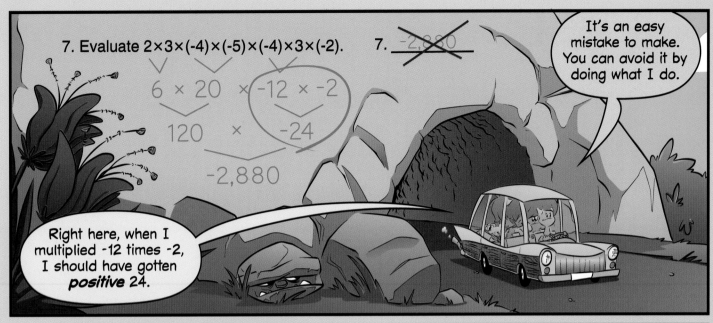

63

Practice: Pages 44-57

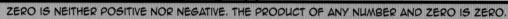

ZERO IS NEITHER POSITIVE NOR NEGATIVE. THE PRODUCT OF ANY NUMBER AND ZERO IS ZERO.

$4 \times 3 = 12$, so $12 \div 3 = 4$

$-4 \times (-3) = 12$, so $12 \div (-3) = -4$

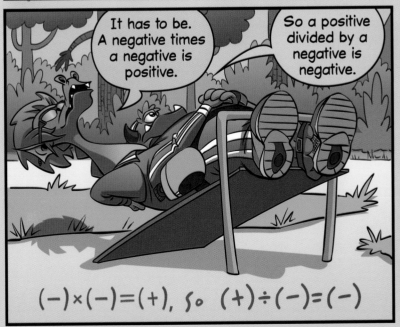

$(-) \times (-) = (+)$, so $(+) \div (-) = (-)$

$-12 \div 3 =$

What is $-12 \div 3$?

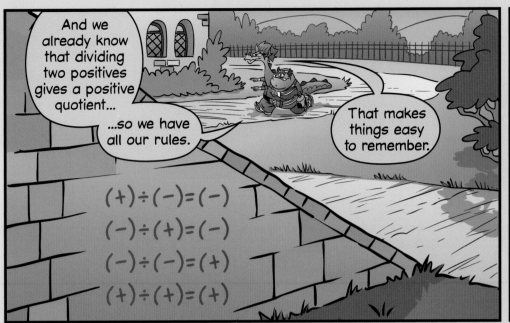

And we already know that dividing two positives gives a positive quotient...

...so we have all our rules.

That makes things easy to remember.

$$(+) \div (-) = (-)$$
$$(-) \div (+) = (-)$$
$$(-) \div (-) = (+)$$
$$(+) \div (+) = (+)$$

What do you mean?

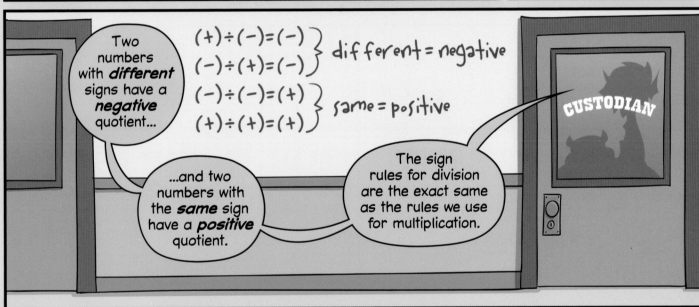

Two numbers with **different** signs have a **negative** quotient...

$$(+) \div (-) = (-)$$
$$(-) \div (+) = (-)$$

different = negative

$$(-) \div (-) = (+)$$
$$(+) \div (+) = (+)$$

same = positive

...and two numbers with the **same** sign have a **positive** quotient.

The sign rules for division are the exact same as the rules we use for multiplication.

CUSTODIAN

We have the same sign!

You mean we're both positive?

No, we're both Capricorns!

December

Writing a negative in front of an expression means to take its opposite.

For example, -(3+4) is the opposite of (3+4).

Since 3+4=7, -(3+4)=-7.

RSENIO HALL

-(3+4) = -7

What is the value of -(-10)?

-(-10)

-(-10) is the opposite of -10.

So, -(-10) is 10!

-(-10) = 10

Taking the opposite of a number is the same as multiplying by -1.

We could compute -(-10) as -1×(-10), which equals 10!

-(-10)
= -1×(-10)
= 10

Quite right!

Finding a number's opposite is the same as multiplying the number by -1.

Try simplifying each of these expressions.

$-(5^2) =$

$(-5)^2 =$

$-5^2 =$

$-n$ AND $-1 \times n$ ARE EQUAL FOR ALL VALUES OF n.

$-(5^2) = -(25) = -25$

-(5²) is the opposite of 5². Since 5²=25, -(5²)=-25.

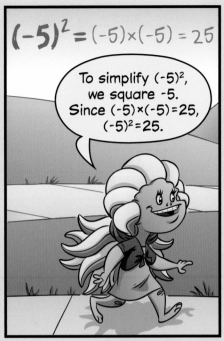

$(-5)^2 = (-5) \times (-5) = 25$

To simplify (-5)², we square -5. Since (-5)×(-5)=25, (-5)²=25.

$-5^2 =$

I'm not sure what -5² means. Do we square -5, or find the opposite of 5²?

ARSENIO HALL

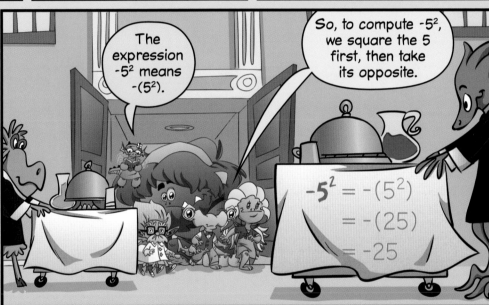

The expression -5² means -(5²).

So, to compute -5², we square the 5 first, then take its opposite.

$$-5^2 = -(5^2)$$
$$= -(25)$$
$$= -25$$

Excellent!

Try a few more.

$-3^4 =$

$(-3)^4 =$

$-(-3^5) =$

Try all 3.

69

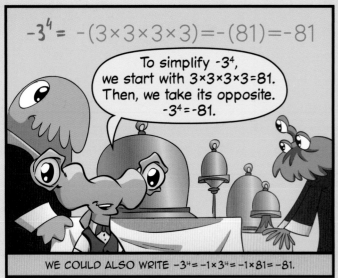

$-3^4 = -(3 \times 3 \times 3 \times 3) = -(81) = -81$

To simplify -3^4, we start with $3 \times 3 \times 3 \times 3 = 81$. Then, we take its opposite. $-3^4 = -81$.

WE COULD ALSO WRITE $-3^4 = -1 \times 3^4 = -1 \times 81 = -81$.

$(-3)^4 = (-3) \times (-3) \times (-3) \times (-3)$

$= 9 \times 9$

$= 81$

To simplify $(-3)^4$...

...we multiply four (-3)'s: $(-3) \times (-3) \times (-3) \times (-3)$. That gives us $9 \times 9 = 81$.

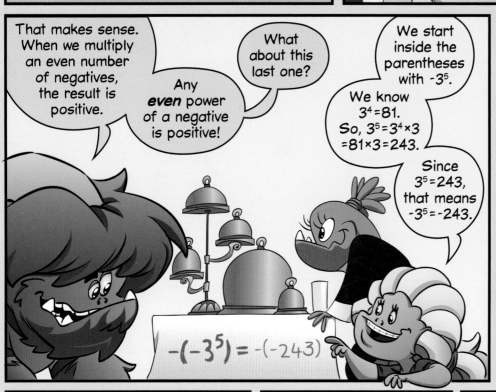

That makes sense. When we multiply an even number of negatives, the result is positive.

Any *even* power of a negative is positive!

What about this last one?

We start inside the parentheses with -3^5. We know $3^4 = 81$. So, $3^5 = 3^4 \times 3 = 81 \times 3 = 243$.

Since $3^5 = 243$, that means $-3^5 = -243$.

$-(-3^5) = -(-243)$

$-(-3^5) = -(-243)$

$= 243$

And the opposite of -243 is 243. So, $-(-243) = 243$.

Exquisite work, little monsters.

Wait, watch out--

Krash!

oof!

Bwah Hah Hah! Professor Grok is gone! I've abducted your educator! It's time for something much more diabolically difficult!

Evaluating such elementary exponential expressions is unconditionally unremarkable.

But simplification of exceedingly elaborate expressions requires **exceptional expertise.**

Compute the value of this expression and relay your response to the register in the refectory.

Refectory?

Muzzle it! I'm monologuing.

Answer correctly, and your esteemed educator will be emancipated.

But answer incorrectly, and he will be *perpetually impounded* in *imponderable peril!*

$$\Big((-2)^{100} \div (-2^{99})\Big) + \Big((-2^{98}) \div (-2)^{97}\Big)$$

Wow, this looks complicated!

How should we start?

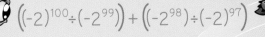

$$\Big((-2)^{100} \div (-2^{99})\Big) + \Big((-2^{98}) \div (-2)^{97}\Big)$$

Can you simplify this expression?

71

IN CASE YOU WERE WONDERING, 2^{100} = 1,267,650,600,228,229,401,496,703,205,376.

Let's simplify the right side.

-2⁹⁸ is negative.

Then, raising -2 to any odd power is negative, so $(-2)^{97}$ is negative.

$(-2)^{97} = (-2^{97})$.

So, on the right side of the expression, we are dividing two negatives.

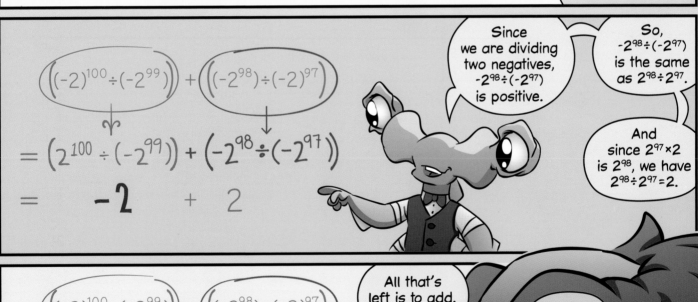

Since we are dividing two negatives, $-2^{98} \div (-2^{97})$ is positive.

So, $-2^{98} \div (-2^{97})$ is the same as $2^{98} \div 2^{97}$.

And since $2^{97} \times 2$ is 2^{98}, we have $2^{98} \div 2^{97} = 2$.

All that's left is to add. -2+2=0!

To the refectory!

What's a refectory?

I thought one of you knew.

A refectory is "a room used for serving meals."

To the cafeteria!

Yum!

MENU

CORN DOGS & TOTS

Okay, I guess I just push 0 on the register?

CLEAR · NO SALE · CAN

7 8 9 · VOUCHER
4 5 6 · SUB TOTAL
1 2 3 · CASH TENDER
. · TOTAL

KLIK

Wooooosh

You've rescued me again, little monsters!

ding!

And just in time for muffins!

Integer Operations Summary:  Lizzie

Multiplying Positives and Negatives:

Same Sign: Positive

$5 \times 7 = 35$

$(-3) \times (-6) = 18$

$8 \times 4 = 32$

$(-7) \times (-2) = 14$

Different Sign: Negative

$-5 \times 7 = -35$

$3 \times (-6) = -18$

$-8 \times 4 = -32$

$7 \times (-2) = -14$

Dividing Positives and Negatives:

Since division is the same as multiplying by the reciprocal, the rules for dividing positives and negatives are the same as the rules for multiplying positives and negatives.

Same Sign: Positive

$35 \div 7 = 5$

$(-18) \div (-6) = 3$

Different Sign: Negative

$-32 \div 4 = -8$

$14 \div (-2) = -7$

> **✳✳✳ SUPER IMPORTANT! ✳✳✳**
> Do NOT try to use the rules for multiplication
> and division for addition and subtraction.

Adding Positives and Negatives:

On the number line, start at the first number.
To add a positive, move right. To add a negative, move left.

$7 + (-9) = -2$ $5 + 11 = 16$

$-7 + 9 = 2$ $-5 + (-11) = -16$

Subtracting Positives and Negatives:

To subtract a number, you can add its opposite.

$6 - (-7) = 6 + (7) = 13$ $-4 - (-3) = -4 + (3) = -1$

$-5 - 9 = -5 + (-9) = -14$ $7 - 18 = 7 + (-18) = -11$

(Sometimes, it's better to leave subtraction as subtraction.)

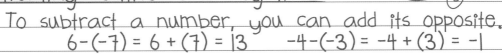

Exponents:

-5^4 means $-(5^4)$, which is $-(5 \times 5 \times 5 \times 5) = -(625) = -625$.

$(-5)^4$ is $(-5) \times (-5) \times (-5) \times (-5) = 625$.

Contents: Chapter 3

See page 70 in the Practice book for a recommended reading/practice sequence for Chapter 3.

Ten-hut!

Until now, you little monsters have only been using the most basic math symbols.

Today, we take off our training wheels and learn to write like true math beasts!

Hammerhead, how would you write "three times x"?

That multiplication symbol looks just like an x!

Sir, like this, sir?

$3 \times x$

In the heat of mathematical computation, we can't afford that kind of confusion!

Starting today, you will replace the rudimentary multiplication symbol with a dot.

A dot?

That's right, Grape Ape.

From now on, the symbol we use for multiplication will be a dot.

$2 \cdot 5 = 10$

$7 \cdot 6 = 42$

$8 \cdot 4 = 32$

$3 \cdot 9 = 27$

The dot is just like this VR-960A hoversphere...

...it's quick, efficient, and looks nothing like a variable.

Sometimes, we ditch the dot entirely.

$$3 \cdot x = 3x$$

The product of a number and a variable is written without a dot.

"Three times x" is written "$3x$."

WHEN WE WRITE THE PRODUCT OF A NUMBER AND A VARIABLE, THE NUMBER IS WRITTEN BEFORE THE VARIABLE. ($3x$, NOT $x3$.)

$$4 \cdot (\text{-}5) = 4(\text{-}5) = \text{-}20$$
$$9 \cdot (\text{-}7) = 9(\text{-}7) = \text{-}63$$

For most products, the dot is unnecessary.

We can use parentheses to represent multiplication.

THE EXPRESSIONS $9 \times (\text{-}7)$, $9 \cdot (\text{-}7)$, $9(\text{-}7)$, AND $(9)(\text{-}7)$ ALL MEAN "9 TIMES NEGATIVE 7."

These expressions may look weird and unfamiliar to you now, but with practice you will learn to use dots and parentheses to write multiplication.

Time to drill!

Evaluate these four expressions.

$$\text{-}9 \cdot 2$$

$$\text{-}7(\text{-}6)$$

$$8(6-9)$$

$$(3+4)(1-5)$$

Try all four.

79

The dot between -9 and 2 means we multiply -9 times 2.

$-9 \cdot 2$
$= -18$

-9 · 2 = -**18**.

Here, we multiply -7 times -6. The product of two negatives is positive.

So, -7 times -6 is **42**.

$-7(-6)$
$= 42$

Since 6−9 is in parentheses, we subtract first. 6−9=-3.

Then, 8 times -3 is -24.

$8(6-9)$
$= 8(-3)$
$= -24$

And for this last one, we evaluate what's in both pairs of parentheses.

$(3+4)(1-5)$
$= (7)(-4)$
$= -28$

Then, we multiply (7)(-4) =-28.

Good work, polliwogs.

Next, not only will your multiplication expressions look different from now on...

...we will also phase out this symbol for division.

Sir, what symbol will replace the division symbol, sir?!

It is a symbol you are familiar with already.

From now on, we will use a fraction bar.

Pinkie! How do we write 45÷9 using a fraction bar?

$$45 \div 9 =$$

Sir, 45 over 9, sir!

$$45 \div 9 = \frac{45}{9}$$

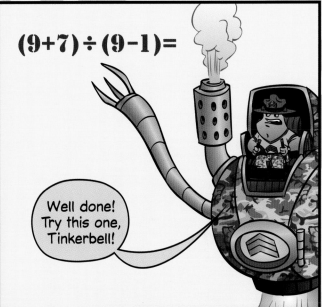

$$(9+7) \div (9-1) =$$

Well done! Try this one, Tinkerbell!

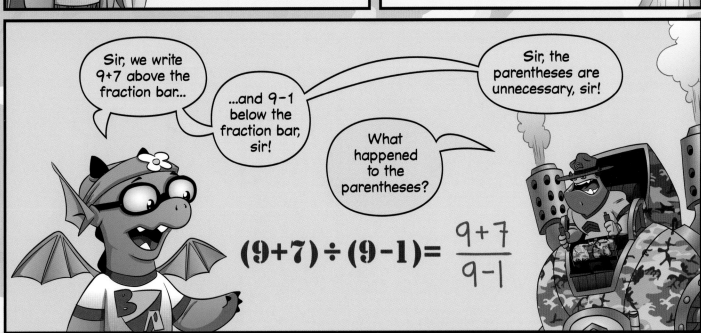

Sir, we write 9+7 above the fraction bar...

...and 9−1 below the fraction bar, sir!

Sir, the parentheses are unnecessary, sir!

What happened to the parentheses?

$$(9+7) \div (9-1) = \frac{9+7}{9-1}$$

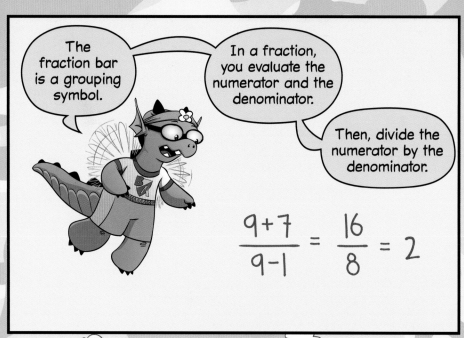

The fraction bar is a grouping symbol.

In a fraction, you evaluate the numerator and the denominator.

Then, divide the numerator by the denominator.

$$\frac{9+7}{9-1} = \frac{16}{8} = 2$$

It's like having parentheses around the whole numerator and the whole denominator.

$$\frac{(9+7)}{(9-1)}$$

Well done, tadpoles.

That's enough introductory expressions.

It's time to put on your big-beast pants and start computing more operations in a single expression.

You must be able to do so with speed and accuracy.

How do we evaluate this expression?

$$\frac{3(5-7)^2}{-2 \cdot 3} =$$

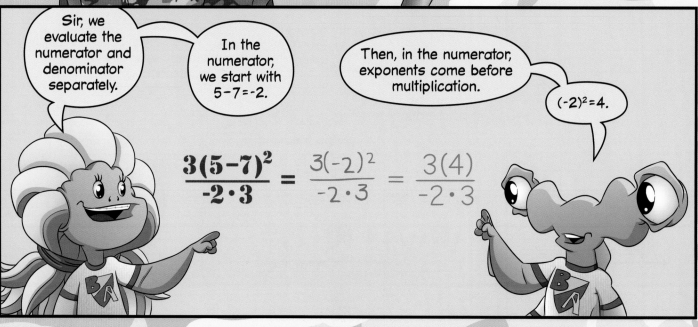

Sir, we evaluate the numerator and denominator separately.

In the numerator, we start with $5-7=-2$.

Then, in the numerator, exponents come before multiplication.

$(-2)^2 = 4$.

$$\frac{3(5-7)^2}{-2 \cdot 3} = \frac{3(-2)^2}{-2 \cdot 3} = \frac{3(4)}{-2 \cdot 3}$$

Now, we can compute the numerator and the denominator.

3 times 4 is 12, and -2 times 3 is -6.

$$\frac{3(5-7)^2}{-2 \cdot 3} = \frac{3(-2)^2}{-2 \cdot 3} = \frac{3(4)}{-2 \cdot 3} = \frac{12}{-6} = -2$$

And to finish, we divide 12 by -6 to get -2, sir!

That's right.

Your math skills are in peak condition.

But, your physical skills need work before this weekend's outdoor retreat.

It's time for physical fitness drills!

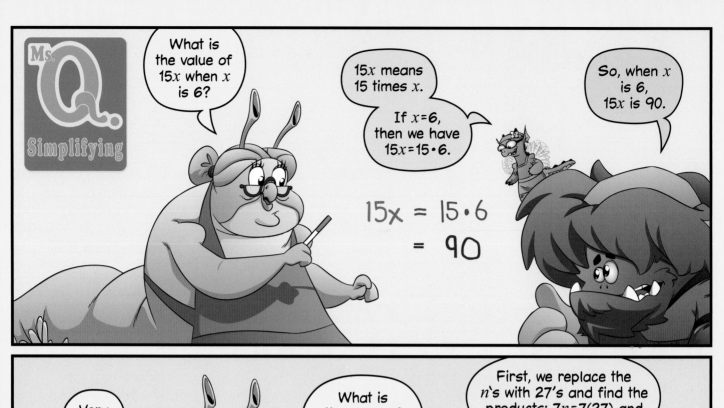

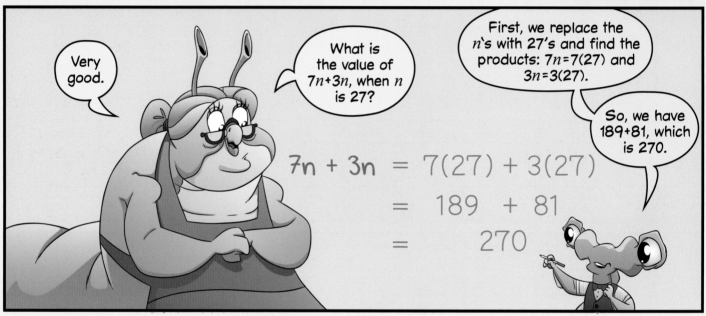

We can use the distributive property.

We factor the *n* out of each term to get $7n+3n=(7+3)n$...

...which is 10*n*.

$$7n + 3n = (7+3)n$$
$$= 10n$$

That makes sense, because $7n=n+n+n+n+n+n+n$, and $3n=n+n+n$.

So, $7n+3n = (n+n+n+n+n+n+n)+(n+n+n)$.

That makes 10*n*.

$$7n + 3n = (n+n+n+n+n+n+n)+(n+n+n)$$
$$= 10n$$

Wait... Why did we do that?

Because computing 10*n* when *n* is 27 is easy!

10 times 27 is 270.

$$7n + 3n = (7+3)n$$
$$= 10n$$
$$= 10(27)$$
$$= 270$$

Very good. Simplifying expressions can be very useful.

How would you simplify each of these expressions?

$$5a - 3a$$
$$x + 2x + 3x$$
$$2j + 3k + 5j - k$$

Try all three.

85

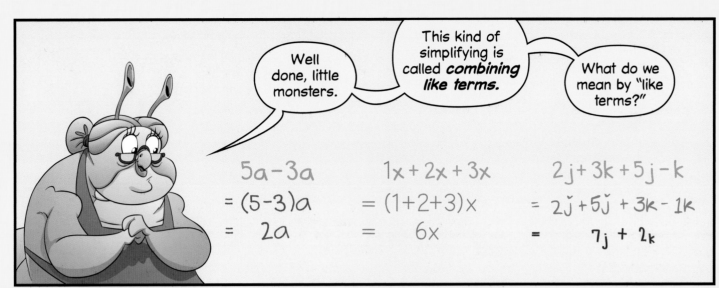

First, we put the *e*'s together and the *f*'s together. We can change all the subtraction to addition, then rearrange.

That way, the like terms are together.

$$7f + 3 - 5e - 4f + 9e + 10$$
$$= 7f + 3 + (-5e) + (-4f) + 9e + 10$$
$$= 7f + (-4f) + (-5e) + 9e + 3 + 10$$

Then we add the *e*'s, the *f*'s, and the numbers separately.

$$7f + 3 - 5e - 4f + 9e + 10$$
$$= 7f + 3 + (-5e) + (-4f) + 9e + 10$$
$$= 7f + (-4f) + (-5e) + 9e + 3 + 10$$
$$= 3f + 4e + 13$$

Then, since *e*=12 and *f*=13, we have 3(13)+4(12)+13.

$$7f + 3 - 5e - 4f + 9e + 10$$
$$= 7f + 3 + (-5e) + (-4f) + 9e + 10$$
$$= 7f + (-4f) + (-5e) + 9e + 3 + 10$$
$$= 3f + 4e + 13$$
$$= 3(13) + 4(12) + 13$$

That gives us 39+48+13, which is 100!

$$7f + 3 - 5e - 4f + 9e + 10$$
$$= 7f + 3 + (-5e) + (-4f) + 9e + 10$$
$$= 7f + (-4f) + (-5e) + 9e + 3 + 10$$
$$= 3f + 4e + 13$$
$$= 3(13) + 4(12) + 13$$
$$= 39 + 48 + 13$$
$$= 100$$

Adding math terms is like adding other stuff.

You only add items that are alike.

For example, if I put three more markers in this cup of markers and scissors...

...it only increases the number of markers in the cup, not the number of scissors.

Whoa! Ms. Q., is this a picture of you and Alex Trebeast?

It is! Do you like his show?

My family watches it all the time.

I bet you were an amazing contestant. Did you win?

What do **you** think?

FAMOUS VRAPPERS

POTPOURRI

SWORDS

POTENT POTABLE

$200 $200 $200 $200

$400 $400 $400 $400

$600 $600 $600

$800

$1

What is a sabre?

FOOD THAT STARTS WITH "Q"

What is a quince?

$200

What was the Great Unicorn Rebellion of '49?

$230,872

Who are the three blind yetis?

$0

Who are three monsters who've never been in my kitchen?

Practice: Pages 71-84

R & G

Isolating x

We spend a lot of time *undoing* things.

What do you mean?

We unclog clogs...

...unlock locks...

...unpack packages...

WRINGER PRO 3000

...unspill spills...

...untangle tangles...

...ungrow growth...

BZZZZZ

...un--

Okay, I see your point.

Math beasts spend a lot of time undoing things, too.

How so?

1. Divide my number by 5.
2. Add 7 to the result to get 15.

   What is my number?

2. Add 7 to the result to get 15.

$\square + 7 = 15$
$15 - 7 = \boxed{8}$

To find the number that you *added* 7 to to get 15, we *subtract 7* from 15. $15 - 7 = 8$.

1. Divide my number by 5.

$\square \div 5 = 8$

That means that when you *divided* your number by 5, you got 8.

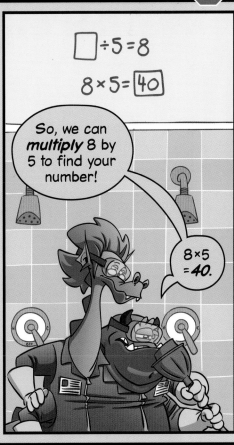

$\square \div 5 = 8$
$8 \times 5 = \boxed{40}$

So, we can *multiply* 8 by 5 to find your number!

$8 \times 5 = 40.$

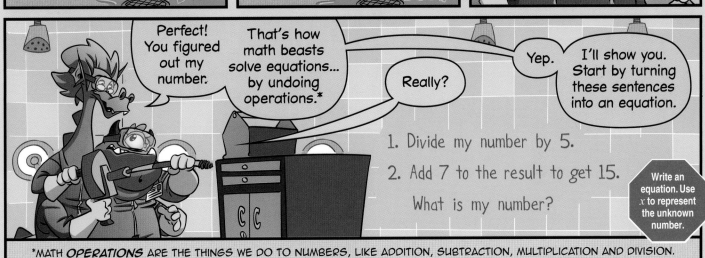

Perfect! You figured out my number.

That's how math beasts solve equations... by undoing operations.*

Really?

Yep.

I'll show you. Start by turning these sentences into an equation.

1. Divide my number by 5.
2. Add 7 to the result to get 15.

   What is my number?

Write an equation. Use *x* to represent the unknown number.

*MATH **OPERATIONS** ARE THE THINGS WE DO TO NUMBERS, LIKE ADDITION, SUBTRACTION, MULTIPLICATION AND DIVISION.

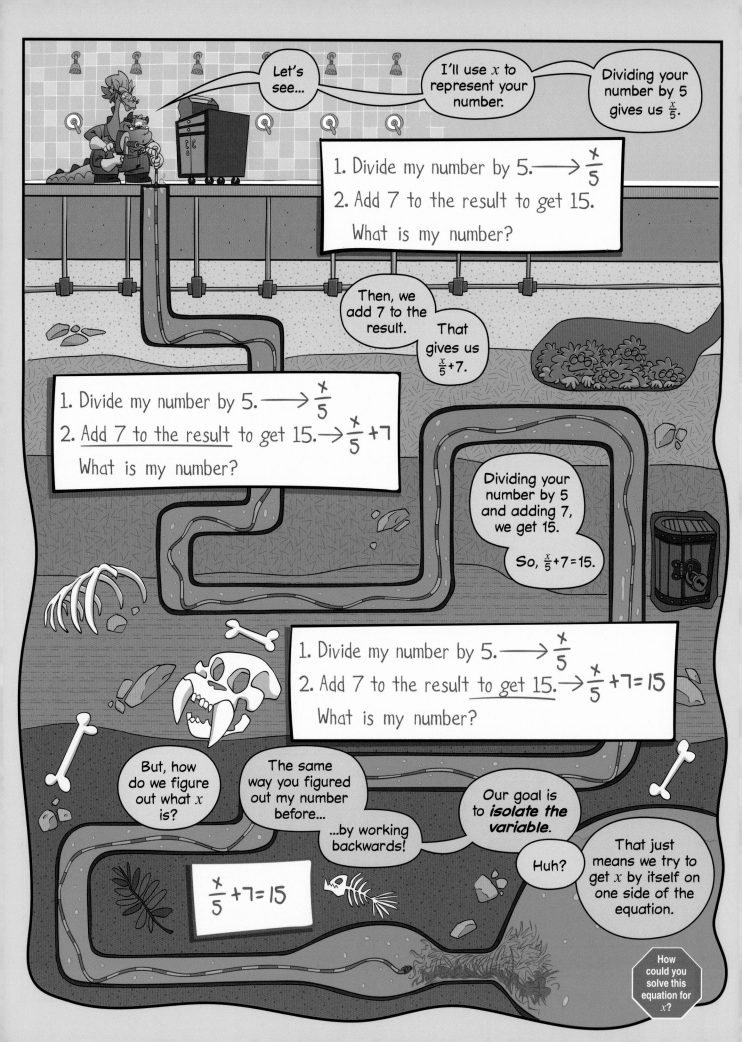

REVIEW THE VARIABLES CHAPTER OF BEAST ACADEMY 3C FOR AN INTRODUCTION TO SOLVING EQUATIONS.

How could you finish solving the equation?

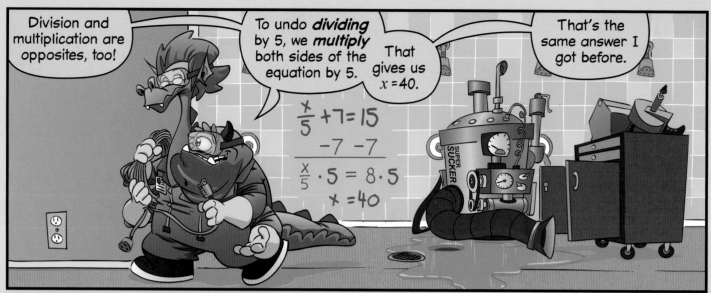

You'll be expected to know how to solve all sorts of equations in tomorrow's Math Meet against the bots.

The bots are incredibly fast at this sort of thing, so you'll need lots of practice to have any chance at keeping up.

Start by solving for x in each of these equations.

$$3x + 1 = 7$$
$$\frac{x}{4} - 6 = 5$$
$$5(x - 6) = 15$$

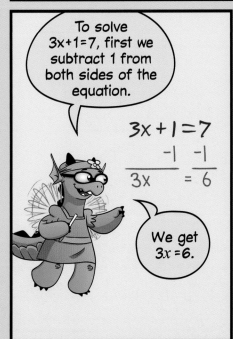

To solve $3x+1=7$, first we subtract 1 from both sides of the equation.

$$3x + 1 = 7$$
$$\underline{-1 \quad -1}$$
$$3x \quad = 6$$

We get $3x = 6$.

Then, we divide both sides by 3 to get $x = 2$.

$$3x + 1 = 7$$
$$\underline{-1 \quad -1}$$
$$\frac{3x}{3} = \frac{6}{3}$$
$$x \quad = 2$$

We can check our work by plugging in 2 for x.

$$3x + 1 = 7$$
$$3(2) + 1 = 7$$
$$6 + 1 = 7$$
$$7 = 7 \checkmark$$

Solve the other two equations.

We solve this equation by adding 6 to both sides...

$$\frac{x}{4} - 6 = 5$$
$$\underline{+6 \quad +6}$$
$$\frac{x}{4} \qquad = 11$$

$$\frac{x}{4} - 6 = 5$$
$$\underline{+6 \quad +6}$$
$$\frac{x}{4} \cdot 4 = 11 \cdot 4$$

...then multiplying both sides by 4.

$$\frac{x}{4} - 6 = 5$$
$$\underline{+6 \quad +6}$$
$$\frac{x}{4} \cdot 4 = 11 \cdot 4$$
$$x = 44$$

So, $x = 44$.

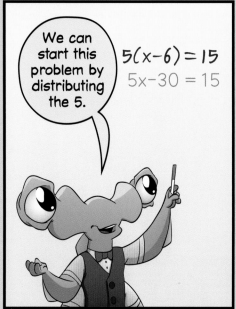

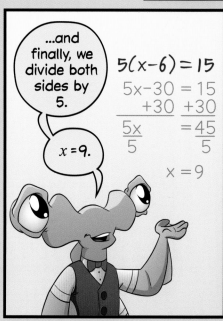

We can start this problem by distributing the 5.

$$5(x-6) = 15$$
$$5x - 30 = 15$$

Then, we add 30 to both sides...

$$5(x-6) = 15$$
$$5x - 30 = 15$$
$$\underline{+30 \quad +30}$$
$$5x \quad = 45$$

...and finally, we divide both sides by 5.

$x = 9.$

$$5(x-6) = 15$$
$$5x - 30 = 15$$
$$\underline{+30 \quad +30}$$
$$\frac{5x}{5} \quad \frac{=45}{5}$$
$$x = 9$$

Neat.

What, Winnie?

I got the same answer without distributing the 5.

How?

How else could you solve $5(x-6)=15$?

96

Multiplication and division undo each other.

$5(x-6)=15$

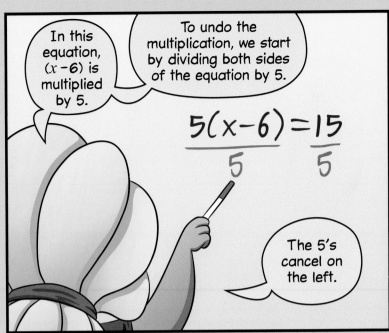

In this equation, $(x-6)$ is multiplied by 5.

To undo the multiplication, we start by dividing both sides of the equation by 5.

$$\frac{5(x-6)}{5}=\frac{15}{5}$$

The 5's cancel on the left.

Dividing $5(x-6)$ by 5 gives us $(x-6)$.

$$\frac{5(x-6)}{5}=\frac{15}{5}$$
$$x-6 = 3$$

So, we have $x-6=3$.

Then, I added 6 to both sides to get $x=9$.

$$\frac{5(x-6)}{5}=\frac{15}{5}$$
$$x-6 = 3$$
$$+6 \quad +6$$
$$x \quad = 9$$

Hmmm... I tried something else that **didn't** work.

$$5(x-6)=15$$
$$+6 \quad +6$$
$$5(x) \quad =21$$

I started by adding 6 to both sides.

Why doesn't the plus 6 undo the minus 6?

Why not?

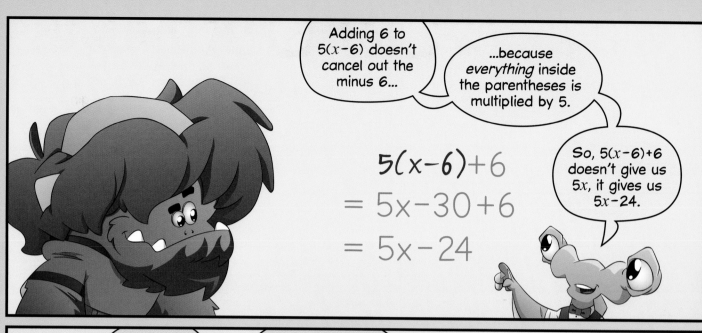

Adding 6 to 5(*x*−6) doesn't cancel out the minus 6...

...because *everything* inside the parentheses is multiplied by 5.

$$5(x-6)+6$$
$$= 5x-30+6$$
$$= 5x-24$$

So, 5(*x*−6)+6 doesn't give us 5*x*, it gives us 5*x*−24.

There are often several different ways to solve an equation.

It's usually best to consider what happens to *x*, and in what order.

Then, we work to undo those steps one at a time to isolate the variable.

Sometimes, isolating the variable can be a little tricky.

We can't always isolate the variable just by undoing operations.

Brrriiing!

Uh oh! We're out of time.

You'll have to try this last one for homework.

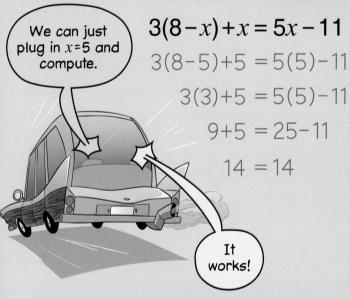

The Little Monsters rang in first.

30!

I'm sorry, that is incorrect.

44.

44 is correct. The Bots get the first point.

Oh no! I subtracted 7 first.

Then, I multiplied by 3.

I should have multiplied both sides by 3, **then** subtracted.

That's ok. None of us could have beaten the Bots to that answer.

$$\frac{x+7 \,-7}{3} = 17 -7$$

$$\frac{x}{3} \cdot 3 = 10 \cdot 3$$

$$x = 30$$

$$\frac{x+7}{3} \cdot 3 = 17 \cdot 3$$

$$x+7 = 51$$

$$\frac{-7 \qquad -7}{x \quad = 44} \checkmark$$

BECAUSE ALL OF $x+7$ IS DIVIDED BY 3, WE CANNOT SUBTRACT 7 FROM $\frac{x+7}{3}$ BY SIMPLY SUBTRACTING 7 FROM THE NUMERATOR. $\frac{x+7}{3} - 7$ IS NOT EQUAL TO $\frac{x}{3}$.

Question 2: Seven times the sum of x and 4 is 36. Express x as a mixed number in simplest form.

It looks like the Bots are arguing. Let's write an equation.

Is the equation $7x + 4 = 36$, or $7(x+4) = 36$?

Which equation is correct?

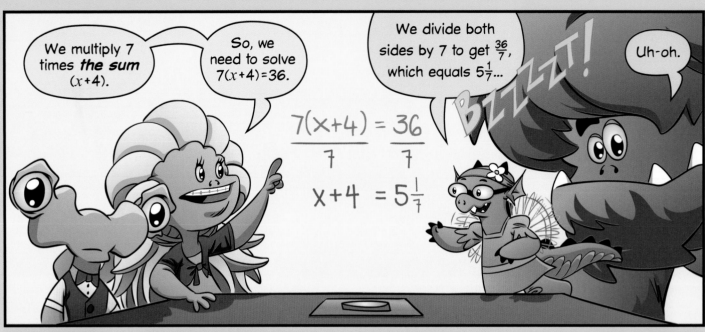

THE TALL BARS IN THE EXPRESSION |−2x+5| INDICATE ABSOLUTE VALUE, WHICH IS A NUMBER'S DISTANCE FROM ZERO. FOR EXAMPLE, |−7|=7 AND |3|=3. LEARN MORE ABOUT ABSOLUTE VALUE IN THE INTEGERS CHAPTER OF BEAST ACADEMY 4C.

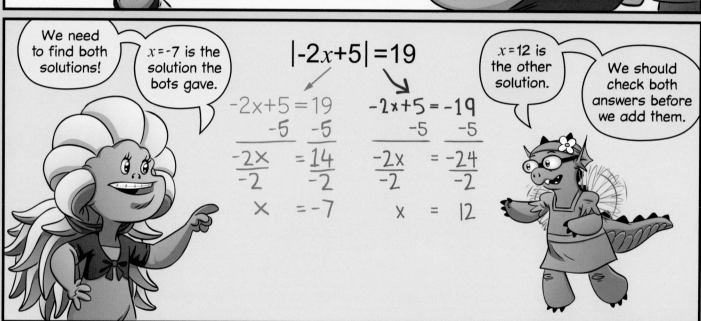

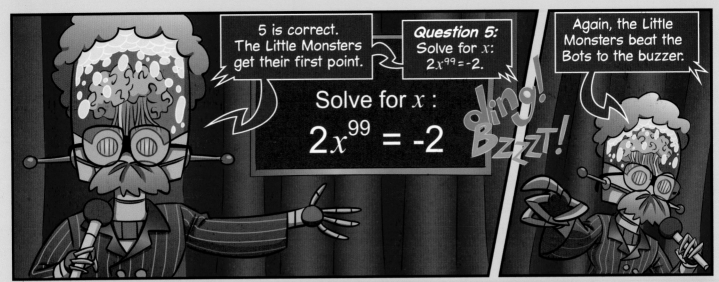

The last question will determine the winner of the Math Meet.

Question 6:
The formula used to compute the volume of a cone is:

$$V = \frac{B \cdot h}{3}$$

V is the volume of the cone. B is the area of the cone's base. And h is the height of the cone.

Find the height of a cone whose base has an area of 7 square centimeters and whose volume is 20 cubic centimeters. Express your answer as a mixed number.

Try it.

Weird. It looks like the Bots are trying to find the radius of the cone's base.

That could give the team the time they need to win!

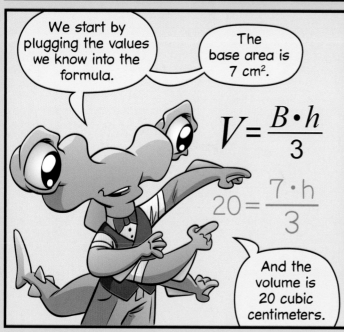

We start by plugging the values we know into the formula.

The base area is 7 cm².

And the volume is 20 cubic centimeters.

$$V = \frac{B \cdot h}{3}$$

$$20 = \frac{7 \cdot h}{3}$$

We multiply both sides by 3 to get $60 = 7 \cdot h$.

$$V = \frac{B \cdot h}{3}$$

$$20 \cdot 3 = \frac{7 \cdot h}{3} \cdot 3$$

$$60 = 7 \cdot h$$

Then, we divide both sides by 7.

$h = \frac{60}{7}$.

$$V = \frac{B \cdot h}{3}$$

$$20 \cdot 3 = \frac{7 \cdot h}{3} \cdot 3$$

$$\frac{60}{7} = \frac{7 \cdot h}{7}$$

$$\frac{60}{7} = h$$

We convert to a mixed number...

Index

For additional books,
printables, and more, visit

BeastAcademy.com